YOUR KNOWLEDGE HAS VALUE

- We will publish your bachelor's and master's thesis, essays and papers

- Your own eBook and book - sold worldwide in all relevant shops

- Earn money with each sale

Upload your text at www.GRIN.com and publish for free

Bibliographic information published by the German National Library:

The German National Library lists this publication in the National Bibliography; detailed bibliographic data are available on the Internet at http://dnb.dnb.de .

Imprint:

Copyright © 2016 GRIN Verlag, Open Publishing GmbH
Print and binding: Books on Demand GmbH, Norderstedt Germany
ISBN: 9783668431515

Samuel Ekwu

Snail Farming. Risk Factors, Diseases and Conservation Practice in the Humid Tropics

GRIN Publishing

GRIN - Your knowledge has value

Since its foundation in 1998, GRIN has specialized in publishing academic texts by students, college teachers and other academics as e-book and printed book. The website www.grin.com is an ideal platform for presenting term papers, final papers, scientific essays, dissertations and specialist books.

Visit us on the internet:

http://www.grin.com/

http://www.facebook.com/grincom

http://www.twitter.com/grin_com

Snail Farming: Risk Factors, Diseases and Conservation Practice in the Humid Tropics

Author : **EKWU, UGOCHUKWU SAMUEL**

[1]Department of Animal Science, University of Nigeria, Nsukka, Enugu State.

Abstract

In recent times, the wild snail population in West Africa region has witnessed a steady decline in population and biodiversity attributed to the impact of human activities, predators, climatic factors and diseases. With the expected 30% rise in the world's population from 7.03 billion in 2010 to 9.14 billion in 2030, adequate measures should be taken and adopted to ensure the continued existence of these economic snails biodiversity in the ecosystem. This paper examines the various human and non human factors endangering snail species population, growth, development and reproduction both in their wild and under domestication; it also highlight the diseases affecting edible snails, conservation consciousness and preventive management practices to be adopted by snail farmers in West Africa in order to ensure the continued existence of these indigenous snail species of West Africa origin with its enormous nutritional, health and economic benefit to man.

Keywords: Snail Farming, Snail Risk Factors and Diseases, Snail Predators, Snail Species Conservation, Humid Tropics

Introduction

Snails have the highest number of recorded extinction known in recent times; many more edible snail species are under threat of extinction. Akinfemi et al., (2014) stated that the rainforest and swamp forest zones of Nigeria are snail's natural habitat. This endangering situation of the snail's species is as a result of the societal behavior of the rural family dwellers. It is a common practice by women and children in rural communities with good forest vegetation suitable for snail habitation in their wild, to go into the forest when it rains to gather, hunt and hand pick snails for food and to sell in the rural markets to earn income for family expenses (Ba, 1994; Agbogidi et al., 2008).This has contributed to the depletion of snail natural stock. It is an established fact that the bulk of the snails meat consumed are obtained from their wild before they reach maturity (Esak and Takerhash, 1992), this has resulted in the depletion and decline of the wild snail species population at a faster rate. Snail meat is a relished delicacy found in the diets of people living in the southern rainforest and the riparian northern guinea savannah zones of Nigeria (Ibom, 2009). Many factors are attributed to this including high species demands, and a significant consumption increase. The need to have snail meat coming from organized rearing system and breeding, in order to achieve a substantial and sustainable snail meat supply to the teeming population in Nigeria becomes necessary (Ebenebe, 2000; Ariry, 2014). According to Agbogidi et al., (2008) the need for increased in animal protein consumption of the rural and urban Nigerian populace in the face of inflation has resulted in the increase in the consumption of snails as food and as a source of income to the peasant farmers and families in rural areas.

Some of the constraints to snails and snail farming are lack of foundation stocks, lack of technical knowledge of snail farming and management, seasonality of snail feeds and vegetation, Snail habitat destruction, climatic and environmental factors, diseases and parasites, predators, genetic constituents and human population growth.

Risk Factors and Human Activities that affect African Land Snails Population

Human activities constitute the largest threat, followed by climatic conditions that affect the snails' biodiversity and population of snails in a given region. They include

- Deforestation i.e. the falling of trees for various purposes such as urbanization, road construction, building of industries, houses and schools.
- Use of pesticides, fertilizers, nematocides by crop farmers in weeds and nematode control in the farming. The use of agrochemicals such as pesticides in crop production also contributed in the change in the soil pH and the snail's ecosystem and biodiversity.

- Slash and Burn agricultural practices of rural farmers during planting seasons. This displaces the snails in their natural habitat and exposes their eggs to harsh weather conditions such as sunlight and rain.

- Bush burning/ fires occasioned by the hunters of wild life animals e.g. bush meat
- Population growth and demand for cheap sources of animal protein
- Indiscriminate snail hunting in the forest (Ikojo et al., 2014).
- Lack of suitable foundation stock for large scale and commercial snail farming
- Unavailability of commercial snail feeds and concentrates

Commercial and intensive crop production systems, which uses inorganic fertilizers as a source of soil nutrients supplementation for profitable crop production. The use of inorganic fertilizers and compounds has destroyed the natural snail ecosystem and habitat suitable for snail growth, development and reproduction in their wild (Ikojo et al., 2014).

Climatic Factors Affecting African Land Snail Population in the Humid Tropics

Unfavorable climatic conditions and climate change poses a serious problem to snail population and farming in West African region. Climatic influence and its effects on snails have an associative effect on temperature, humidity, wind/air movement and light intensity. Snails are susceptible to infections and diseases caused by environmental contaminations and pollutants (Ebenso and Ologhobo, 2009).

Temperature: It is a major factor that influences the activities of snails. Snail requires lower and moderate temperature for normal feeding and body functions. Temperature ranges of between 23 - 28^{oc} are suitable for snails' growth and development. During hotter periods with high ambient environmental temperature, snails may experience heat stress. Heat stress poses a serious economic threat to snails population and snails under domestication. According to Ariry (2014), the effect of heat occasioned by high temperatures on snails could be in the form of reduced feed intake and utilization, reduced egg production, reduce growth rate, low body weight, poor hatchability and fertility. Okafor (2001) stated that snails hibernates and aestivation especially during dry hot unfavorable seasons. Air, environmental and noise pollutions (Ebenso and Ologhobo, 2009) affects snails.

Humidity: Snails enjoy moist and cooler environments, which is usually achieved when it rains, the atmospheric air become moist with high relative humidity. Snails are active at a relative humidity of between 70 - 90%. However acidic rains in some areas caused by high air pollutants adversely affect snails' population (Ojiako, 2006). If the air is drier and hot, for longer periods, snails may dry up and die or hibernate. Snail pens and the top soils should be sprinkle with water regularly during hot periods especially in the dry seasons to maintain adequate damp environment (Ikechukwu, 2012).

Excessive air movement and winds: This is another climatic factor that may result in severe dehydration in snails and may cause them to retract into their shells, rather than feeding to gain weight and to breed. Prolong periods may cause the snails to go into aestivation which is a period of inactivity or dormancy in the life of snails (Okafor 2001).

Light Intensity: Snail requires light for their activities such as feeding and breeding. Though snails are nocturnal animals, however they require light for some photo-biological processes essential in the energy level and food chain e.g. cellular digestion and photosynthesis (Ikechukwu, 2012). Day light is usually from the sun, longer periods of light stimulate and prompt snails into reproduction under favorable conditions.

Soil Type: Snails depend very much on the soil for their food and reproduction e.g. egg laying. They can hardly survive or thrive effectively in the absence of a suitable soil-type. They require moist, aerated, easily drained, non waterlogged and non acidic soils. Soil rich in minerals contents and organic matters are good as too soil for snail farming after being sterilized to kill pathogens in the soils (Ikechukwu, 2012).

Predators of African Land Snails in the Humid Tropics

Giant African land snails are faced with the challenge of predator in their natural wild habitat, these predators pose a great danger to their normal growth and reproduction of snails, if not checkmated may lead to the decline in population of snail biodiversity or extinction of various snail species in nature. Predators mostly depend on their prey as a source of food for survival in the ecosystem. Snail predators feed on the snail species at their various stages of growth and maturity. According to Akinnusi (2014), snail predators includes: *Arthropods* – Insects (termites, beetle, mites, moth, driver ants, carabid beetles, cockroaches and soldier ants); *Crusteceans:* – Millipedes, Centipedes, Cricket, Crabs and Forest Spider; *Reptiles:*- Lizards and Snakes, *Amphibians:*- frogs, turtles and toads; nematodes, *Rodents:*- mice and rats; *Aves:*- birds, crows, ducks and turkeys as well as *Mammal:*- man (Ikechukwu, 2012).

Diseases and Parasite affecting African Land Snails in the Humid Tropics

The common diseases affecting snails either in their wild or in their cultured environments includes:

Fungal diseases mainly *Fusarium Spp* affects indigenous snail species native to West African region. They are susceptible this diseases causing agent. These diseases is commonly referred to as rosy eggs disease and the affect eggs turns reddish brown and die off (Akinnusi, 2014).

Parasites such as *Alluaudihella Flavicornis* are diseases vectors to snails both in their wild and under domestication.

Bacterial diseases caused by *Pseudomonas Spp* especially *Pseudomonas aeruginosa* causes intestinal infections in snails. This disease affects snail's normal growth and development processes.

Deficiency diseases, it occurs mostly in domesticated snails with poor feeding, as a result of lack of minerals nutrients especially calcium and phosphors. The affected snail's shell turns white as a result of deficiency of calcium in their feeds over a longer period of time (Akinnusi, 2014).

Cannibalism: This mostly occurs in domesticated snails housed in pens. Older snails can eat, break the shells or fed on hatchlings as a source of nutrients especially calcium and water to avoid dehydration and for their survival. This occurs where snails are overcrowded and there is increased competition for food and space (Ikechukwu, 2012).

According to Okafor, (2001) and Akinnusi (2014), they stated that predators of snails inflict havoc on the snails by either breaking their shells, biting or sting the snails or eat them as food both their eggs and juvenile snails e.g. frogs and reptiles (Akinnusi, 2014). Snail predators adversely affect the population of the native Giant African Land Snail species of West African origin and its biodiversity.

Generally maintaining of high hygienic standards in snail farms will reduce the incidence of diseases and spread of diseases in the snail farms. According to Walker et al., (1999), snails ingest micro-organisms e.g. bacteria from the soil and their environment, poor hygiene may predispose the snails to diseases and pathogens, which will affect their growth and reproduction.

Preventive Measures under Intensive and Semi-Intensive System of Snail Farming

- Snail housing should be well constructed, secured and strong to withstand and keep out these predators as well as prevent theft by humans.
- Proper covering and netting of snail housing units to avoid and minimize the escape of hatchlings, juvenile or adult snails from their pens.
- Doors that fit well into the wall frames should be used to prevent rodents, shrews, frogs and lizards from gaining entrance into the snailery and snail pens.
- The digging of 2 feet deep trenches or gutter preferably concrete gutter for durability round the snailery house and fill the gutter with a salty water and pour red oil or kerosene, to avoid mosquito breeding in the water or you can use other locally available insect repellants e.g. condemned engine oil can be used. The water in the trenches/ gutter will prevent crawling insects e.g. soldier ants, millipedes and centipedes from reaching the snailery house. Also fencing of the snailery is recommended.
- Sterilization of the top soil before being used as bedding in snail pens, this will eradicate parasites, nematodes and other micro organisms present in the soil. The snail pen top soil should be replaced regularly every three months; this is a good diseases management practice. Avoid the used of soils containing harmful chemicals e.g. lead, mercury and salts as top soils.
- Moisten of the snail pen top soil bedding regularly, but not water logging it, snails like and enjoy humid environments, these condition enhance their bioactivities and development. To reduce moist losses from the snail pens, plantain leaves and

cocoyam leaves can be used to cover as shade, disinfect the leaves with salt and water solution.

- When concrete cages are used, the floor of the snail pens should be cemented before filling the pens with the top soil which is already sterilized mostly loamy soil is filled in the various snail pens of the concrete cage. These prevents the entrance of red ants, termites, which burrow from underground, if not cemented, they may attack the eggs laid by the snails.

- Environmental sanitation should be carried out regularly, to ensure safety of cultured snail species. The surroundings of the snail farm should be kept clean, tidy and disinfected by fumigating the environment to kill micro organisms and predators. Weeds and vegetations around the snail farm should be cut regularly and not be allowed to be bushy.

- Overcrowding should be avoided in snail pens, it leads to stunted growth, diseases infestation, competition for food and space, high mortality rate and decline in productivity (Ikechukwu, 2012). Depending on the snail species and age an appropriate stocking density should be used. Snail diseases spread rapidly when snails population are dense and overcrowded.

- Routine checks, monitoring and close observation of snails in their pens are very important, any abnormality in feeding and behavior observed should be corrected immediately. Constant and regular monitoring of the snails will improve the well-being of the reared snail stock. Sluggish, sick and dead snails should be removed from the snail pens, to avoid the spreading of diseases to healthy snails. It helps you to detect opening in the snail pens from where snails can escape.

- Regular removal of remnant feeds in snail pens and provision of clean water to snails *ad libitum* and serving of fresh feeds to snails encourage their smooth growth and development. Left over snails' feeds not removed in the pens may serve as a pathogenic diseases source to snails under domestication. (Walker et al., 1999) stated that snails ingest bacteria from the soil and their immediate environment.

- Stressing of the snails should be minimized, as this may retard their growth rates, the stress could be inform of sudden touching of the snails as they creep, rasp or feeds and improper handling, vibration or noise among other pollutants all causes stress to the snails (Ikechukwu, 2012).

Recommendations

Awareness should be created and more farmers be encouraged to venture into snail farming owing to it cheap initial startup capital, high market demand, less routine and management demands, less space requirements, nutritional, medicinal and economic importance.

Graduates, school leavers, youths, unemployed persons, physically challenged persons and retirees should go for snail farming and husbandry trainings, to acquire the technique and know–how needed for profitable snail farming business enterprise, and be self employed.

Research institutes and universities should be empowered by governments and international donor agencies to carry out researches on our native and well adopted snail species of the humid tropical environment, with a view to establishing its nutrient requirements, feeding standards, reproductive behavior, biology and physiology of breeding/ reproduction, stocking density and resistant to common endemic diseases as well as best farming system to adopt.

Government should encourage people to go into snailery by subsiding snail farming inputs and the provision of good snail parent / foundation stock to individuals and co-operatives farmers that wants to venture into snail farming.

Animal feed millers should development commercially available snails' feeds to be sold to snail farmers at cheaper rate as one of their feed products e.g. snail starter, snail grower feeds and snail finisher feeds. Production of snail feeds in commercial quantities for snail farmers.

Development of genetically improved and fast growing snail species of tropical environment by snail breeders, research institutes and universities.

Snail associations and non-governmental organizations e.g. Research Network for Giant African Land Snails (NetGALS), should promote, publicize all research findings on snail farming; breeding, stocking rates, diseases and management. They should organize snail farming seminar and trainings to intending farmers on quarterly basis in each state capitals in Nigeria for free.

Conclusion

Snails species indigenous to West African are under threat occasioned by human activities, predators, climatic conditions and diseases infestations, conscious efforts should be made to domesticate these natural endowment, and to rear them in commercial and large scale as way of conservation of the snail species and source of economic development. Various methods and management practices to be used to prevent and ensure the continued existence of snail species of West African origin are emphasized in this paper. In view of this, we believe that this will go a long way in increasing the level of animal protein supply needed by the teeming population of West African states.

References

Agbogidi, O. M., Okonta, B. C. and Ezeani, E. L. (2008). Effects of two edible fruits on the Growth Performance of African Giant Land Snail (*Archachitina marginata*).

Akinfemi, A., Adeyeye, W., Nwodu, J. A., Ikojo, H. A. (2014). Open Pasture Snail Farming: In Micro Livestock Production in the Tropics. Jeolas Press, Owerri, Nigeria. Pp. 135 - 145.

Akinnusi, F.A.O (2014). Snail Production and Management. Tolukoya Print House, Abeokuta, Ogun State, Nigeria.

Ariry, E. C. (2014).Problems and Prospects of Snail Production: A case Study of Owerri North L.G. A. of Imo State, Nigeria. National Diploma Project in Animal Production and Health Technology, Imo State Polytechnic, Umuagwo, Nigeria. 32pp.

Ba, C. (1994). Socio-economic Aspects and Nutritive value of the Giant African Snail Meat from Cote D Ivoire. PhD Thesis, Department of Agricultural Economics, Universite National De Cote De Ivoire.

Ebenebe, C. I. (2000). Mini-livestock Production in Nigeria the Present and Future. Proceedings of the 5th Annual Conference of Animal Science Association of Nigeria, Port Harcourt, Nigeria. September 19 - 22.

Ebenso, I. E. and Ologhobo, A. D. (2009). Effect of Lead Pollution at Industrial contaminated sites on Sentinel *Achatina achatina, Bulletin of Environmental Contamination and Toxicology*, 82 (1): 106 -110.

Esak, K. O and Takerhash, I. S. (1992). Snail Pest and Food. *Malaysian Journal of Economic Agriculture* 59: 359 -367.

Ibom, L. A. (2009). Variations in Reproductive and Growth Performance Traits of White-skinned x Black skinned African Giant Snail Hatchling {*Archachatina marginata (Swainson)*} in Obubra, Nigeria. PhD Thesis: Department of Animal Science, University of Calabar, Calabar, Nigeria. Pp. 166.

Ikechukwu, C. C. (2012). Basic Tips on Snail Farming. Creative Forum Nigeria Ltd, Onitsha, Anambra State.

Ikojo, H. A., Agbasu, C. A., and Ahaotu, E. O. (2014).Gastropods. In: Micro Livestock Production in the Tropics. Jeolas Press, Owerri, Nigeria. Pp. 190-196.

Ojiako, E. O. (2006). An Introduction to Mini-Livestock Husbandry. Ekecy Printers and
Publishers, Asaba, Delta State.

Okafor, F. U. (2001). Edible Land Snails: A Manual of Biological Management and Farming
of Snails. Splendid Publishers, Lagos.

Walker, A. J., Dmglen, A. and Shewy, P. R (1999). Bacteria associated with the Digestive
System of Slug *Deroceras Recticulum* are not required for protein digestion, *Soil
Biology and Biochemistry*, 31: 387-394.